AF229865

TATSUNIVAR LAMBOBI

THE NUMBER STORY

SMALL BOOK ONE

ENGLISH - HAUSA

*Numbers Teach Children
Their Number Names*

written and illustrated by

MISS ANNA

Early Reader Edition of *The Number Story 1*
Bronze Medal Winner, 2016 Wishing Shelf Book Award

Copyright © 2018 by Jieeun Woo
Illustrations © Jieeun Woo

Cover by | Lumpy Publishing
Layout by | Lumpy Publishing
Translated by MuhdNG
Coloring by Jieeun Woo and Maria Mirabella

All rights reserved. No part of this book may be reproduced or transmitted in any form or by any means whatsoever, including photocopying, recording or by any information storage and retrieval system, without written permission from the publisher and/or author: missanna@missannabooks.com.

Library of Congress Control Number: 2018902040

Names: Miss Anna, author.
Title: Number story : numbers teach children their number names / Miss Anna.
Description: Portland, OR: Lumpy Publishing, 2018.
Identifiers: ISBN 978-1-949320-06-0| LCCN 2018902040
Summary: The pictures and rhymes present stories which introduce numbers 0-10.
Subjects: LCSH Numeration—English—Hausa--Pictorial works--Juvenile literature. | BISAC JUVENILE NONFICTION /
Languages: English—Hausa
Classification: LCC QA141.3 .M57 2018 | DDC 513—dc23

Publisher: Lumpy Publishing
Website: www.missannabooks.com
Email: missanna@missannabooks.com

Paperback: ISBN 978-1-949320-06-0
Printed in the U.S.A.   1 3 5 7 9 10 8 6 4 2

Kanaso kakoyi sunayen lambobi?
Kinaso kikoyi sunayen lambobi?

It is very easy and a lot of fun!

Yanada sauƙi sosai,
kuma akwai dadi sosai!

Say-along our little jingle

Muyi wannan wakar tare daku!

starting from Number One!

Zamu fara daga lamba ta daya!

1

ONE   looks like my one finger.

DAYA

Tana kama da dan yatsana.

ONE!
DAYA!

2
TWO   trails a tail.

BIYU

tanada jela.

A TAIL! JELA!

3

THREE   has bumps.

UKU

tanada tudu da kwari.

BUMPY!   TUDU DA KWARI!

4

FOUR   carries a sail.

HUDU

tana kama da jirgin ruwa.

A SAIL!
YADIN JIRGIN RUWA!

5

FIVE   is a racing track.

BIYAR

kamar hanyar tsere.

VROOM
Karan tafiya!

6

S I X curves like a snail.

SHIDA

ta lanƙwashe kamar dodon kodi.

A SNAIL! DODON KODI!

7

BE CAREFUL! IT'S SHARP!

KULA! TANADA TSINI!

8

E I G H T   is rollercoaster rails.

TAKWAS

hanyar jirgin qasa mai lanƙwasa.

YE! YE! YE!
YIPPEE!

NINE   is a bubble on a stick.

TARA

kumface ajikin sanda.

A BUBBLE!

KUMFA!

10

TEN is an eye of a whale.

GOMA

idon mari babban kifi ne,

guda daya.

HELLO!
SANNU!

And
Da kuma

0

ZERO   is an empty pail.

SIFILI

bokiti ne mara komai aciki.

IT'S EMPTY!
BA KOMAI ACIKI!

Thank you for playing with us today.

We had a lot of fun too!

Mungode daka/dakika tayamu yin
wannan wasan yau.

Muma munji dadi sosai!

We are your Number friends,
Zero to Ten,
Who will be here for you~
Mun zama abokanai
Sifili zuwa Goma.
Muna tare dakai/dake akoda yaushe.

Bye-bye now!
See you again soon!
Sai anjima!
Sai mun kara haduwa!

The Numbers are *SINGING* too!

To sing-a-long, look for Miss Anna Number Story
at your favorite music store like iTUNES.

MP3

Numbers 0-10
IDENTIFYING
& COUNTING

Numbers 11-20
& Ordinals
first, second, third...

Numbers 0-100
& Place Values
ones, tens, hundreds...

About Clocks
& Telling Time
hours, minutes, seconds

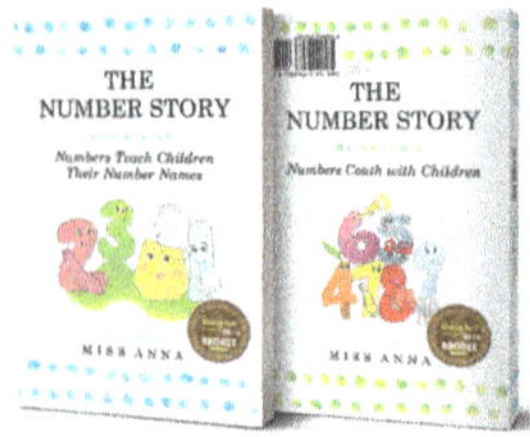

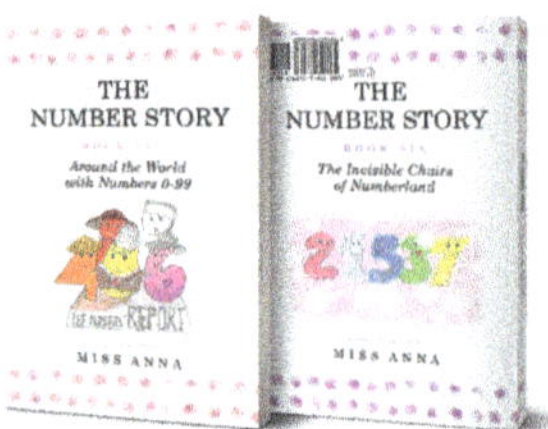

Number Story 1 & 2
isbn: 978-0-996216-48-7

Number Story 3 & 4
isbn: 978-1-945977-01-5

Number Story 5 & 6
isbn: 978-1-945977-06-0

Number Story 7 & 8
isbn: 978-1-949320-40-4

For more Miss Anna books to love,
visit us at

w w w . m i s s a n n a b o o k s . c o m

Numbers are working hard all over the world!
Come Travel the World with Us!

www.ingramcontent.com/pod-product-compliance
Lightning Source LLC
Chambersburg PA
CBHW040900070726
47599CB00035B/2244